History of the Duroc Pig

A Short History of the Duroc Jersey Breed of Swine

by Robt. J. Evans

with an introduction by Jackson Chambers

Self Reliance Books

Get more historic titles on animal and stock breeding, gardening and old
fashioned skills by visiting us at:

http://selfreliancebooks.blogspot.com/

Introduction

I am pleased to present another title in the "Raising Pigs" series..

As with all reprinted books of this age that are intended to perfectly reproduce the original edition, considerable pains and effort had to be undertaken to correct fading and sometimes outright damage to existing proofs of this title. At times, this task is quite monumental, requiring an almost total "rebuilding" of some pages from digital proofs of multiple copies. Despite this, imperfections still sometimes exist in the final proof and may detract from the visual appearance of the text.

I hope you enjoy reading this book as much as I enjoyed re-publishing and making it available to fanciers again.

With Regards,

Jackson Chambers

ROBT. J. EVANS

INTRODUCTION

The most frequent inquiry that has come to my desk each week for the last several years is, "Where can I find a book that contains a history of Durocs?" In order to answer many of the questions that arise in the mind of the student of pedigree, and in the mind of the man who has embarked into Duroc breeding with the intention of mating to produce improvement, I have written this little book. No man can be a breeder of hogs and make headway with his work, and make improvement in the quality and characteristics of his herd animals unless he knows something of the commingling of blood that produced his herd headers. Only a few of the questions that have come to me in the recent years about the breed are answered in this book. Much of the early history of the breed and the early herds has been lost for the want of a man who could devote his time to gathering up the widely scattered bits of information.

I have tried to make the story readable, and in as few words as possible give as much information as could be boiled down in so small a volume. It is by no means a complete history of the breed. The writer hopes to be able to take this larger work up shortly and with the assistance of the leading men who have been active in the business in times past and with those who are leaders now, bring out a book that will give all of the history of the breed, that can be learned by a search of the records, and in conference with those who have through the trying years of the breeds' existence, been working for breed improvement.

This book is dedicated to the interests of the Duroc, a breed of hogs that has come to the front as no other breed has done in the history of livestock production. The Duroc has brought fecundity, stamina and size into the pork herds of the day and a visit to the market centers will prove to the most skeptical that he is the "Farmers and Feeders Favorite." The prices of

Duroc breeding stock at this writing in all sections where purebred hogs are raised is evidence of the popularity of the Duroc, and the reason for that popularity is his ability to make good as a producer of pork under all conditions and environments.

For more than 25 years the writer has worked to scatter the gospel of "More and Better Durocs," and it is with no little pleasure that he sees the ever-widening interest taken in his favorite breed. And this work is put out with the intention and sincere desire to help in still extending the fields of the Duroc.

ROBT. J. EVANS.

HISTORY

of
the

DUROC

IT IS conceded by the best posted men in the Duroc business, both old and new, that there is little that can be relied upon in the early history of the Duroc as really authentic. This one fact, however, is pretty conclusively revealed by the examination of everything that can be had on the subject, that the Duroc-Jersey comes from what was best in two or more strains of red hogs. Among these were the massive, coarse Jersey Reds of New Jersey, which are descendants of a pair of pigs imported from England in 1832; the more compact Duroc of New York, so named by Isaac Frink of that state, after a noted stallion which he owned; and from the Kentucky strain which foundation was imported by the Honorable J. C. Clay who was Minister to Portugal under President Taylor in 1850.

Very naturally, it was some years before there was any attempt to establish a type of any kind. The Jersey-Reds grew to extremely large size when matured, some of the records being as high as 1,053 pounds as barrows, and one record is found of 30 head 22 months old that averaged dressed at 685 pounds. The Duroc branch of the family carried more finish, was more neat in appearance, both in head and body. From all that can be learned of the nature of the J. C. Clay branch, it was a medium type of red hog, and must have originally come from the wild red hog of Northern Africa. Red hogs were found in an early day in every state on the Atlantic coast where slave trading was carried on and it is fair to assume that the original of these came from North Africa.

The Jersey-Reds were named by the Honorable Joseph Lyman in 1857, who was at that time Agricultural Editor of the New York Tribune and the words "Duroc" and Jersey-Reds were used separately for a number of years. The union of these names under the Duroc-Jerseys and the merging into the one breed was made about the time the first steps were taken to organize a recording association which was about 1880.

In Saratoga county, New York, and in Connecticut, and in Vermont, the same type of red hogs was bred and not until 1877 were there any united efforts to agree upon a standard of characteristics and qualifications. This was brought about by the breeders of Saratoga county. A table of characteristics was completed by this county association, and was the work largely of W. M. Holmes, who later moved to Iowa and continued the improvement of the Duroc-Jerseys in his adopted state. This firm was later known as C. H. Holmes & Co., Chas. Holmes of the firm being the first secretary of the American Association, selected in 1883. However, a Wisconsin organization had been formed primarily with the purpose of a record, but had never taken any definite steps towards that end and was not doing any recording.

This organization was formed in 1882 with Geo. A. Lytle of Elkhorn, as president, and W. H. Morris from the same place as secretary.

It was not until 1881-2 and -3 that there was any particular attention paid by the breeders of Duroc-Jerseys to attend the big shows of the country as evidenced by the record in the first volume in the extended pedigrees given in the last part of the volume. These give the show record of these various animals and do not date back earlier than 1881, most of them being in 1882 and 1883. The first animal recorded was Red Jacket No. 1, owned by Thos. Bennett of Illinois.

This animal was a combination of the breeding of Clark Pettit, New York, Samuel Stark of South Bend, C. Burgen, Richmond, Ky., and D. W. Zink of Illinois. It was evident that the breeders were looking everywhere and were buying from different localities to secure the right cross and improve these hogs. We need but cite the pedigree of one of the show hogs of 1884 and 5, Climax 21. He was shown by Railsback & Pittsford of Illinois throughout the West and South. Climax was first at Des Moines in 1884, Omaha 1884, Ottumwa 1884, Oskaloosa the same year, and second and Sweepstakes at Kansas City, second at Sedalia the same year, and first at New Orleans in 1885. This boar was sired by John Jordan 297, the latter being bred by Mr. Bradbury of Nasons, Va., the dam, Tillie (no number), bred by Clark Pettit. In the first volume of the American record will be found the pedigree of animals that were winners in the big fairs from New York to New Jersey, Virginia and as far west as Omaha, Neb. The rapid diffusion of this blood of the Duroc-Jersey throughout the hog belt is evidence that there was need for a better pork-growing machine than found at that time. There is no comparison of the Duroc-Jersey of that day with the improved Duroc of today. Yet he had within his makeup the characteristics that were bound later to make him a leader in swine production. There were strength of character, ruggedness, prolificacy and the ability to put on pounds of pork on forage and concentrated feeds. The Duroc has been developed through more than three quarters of a century of careful consideration for these qualifications, necessary to make the best machine to convert grain and grass into pounds of pork on foot. For the most part he has been developed by men who had to make their living from pork growing. The idea in mind all the way up the line has been the best possible money-making hog.

The development of a breed of livestock comes usually through a few leaders adapted to that kind of work and the improvement of the Duroc is no exception. Early in the history you find the names of Clark Pettit of New York, John S. Collins of New Jersey, Morton of Ohio, Railsback, Browning, Bennett and Stoner of Illinois, Holmes of Iowa, formerly of New York. Ingram of Illinois, Roberts, Stribling and West of Iowa, Harris and Crabb of Kentucky, Searle of Nebraska and Stonebraker of Illinois, all of whom played an important part in the early advancement of the breed.

I fail to find in the first volume of the record, issued by the American, which was the first permanent record of Duroc-Jerseys issued by the breeders of this breed of swine, a name that appears in the recent records of either the American or National records, but when we come to the second volume we find in it the names of many familiar present-day breeders. Only a few, however, whose names are represented in that volume, are still active in the work. The first volume was issued in 1885 and the second in 1890. The National Association was formed in the Fall of 1891 and issued its first volume in October, 1893, the first volumes being delivered at the World's Fair show in Chicago in October, 1893, at a called meeting of the Board, held at that time. At that meeting strong resolutions were passed, copies of which were forwarded to every fair association of the country, asking that a distinct class be made for Durocs. Only in a few of the leading state fairs were these classes made prior to 1893, but the Durocs had been showing against other small breeds at that time.

Connected with the early history of improved and recorded Durocs we find A. Ingram, Pittsford & Railsback, J. M. Browning, G. W. Stoner, Thos. Bennett and C. J. Stuckey in Illinois; Wm. Holmes & Son, J. W. Doak, J. M. Shaw and A. Failor, Iowa; Thos. Lovelock and Wm. Bradbury, Virginia; F. D. Curtis, New York; H. C. Stoll, Nebraska; R. L. Williams, Kentucky; Clark Pettit, New Jersey; P. C. McClure, Samuel Taylor, Ohio; R. H.

Gage and Geo. A. Lytle, Wisconsin. These names appear very often in Volume I of the American record. In the second and third volumes we find the names of J. M. Stonebraker, Jos. Vogel, Illinois; S. E. Morton, Ohio; Amos Harris, Kentucky; Herring & Hummer, Iowa; G. W. Witham, Illinois; John S. Collins, N. J.; I. M. Stansell, Illinois; Rankin & Son, Illinois; William Roberts and J. W. Stribling, Iowa. The leading families of the breed and the hundreds of noted animals produced by their blood lines, with but few exceptions, trace directly to animals recorded in Volumes 1, 2 and 3 of the American record, bred by these breeders whose names we have here listed.

J. M. Stonebraker's first recorded boar was Ben Butler 1387, by Dan Voorhees Jr. 323, by old Dan Voorhees, unrecorded. Dan Voorhees Jr. was used by Railsback & Pittsford of Hopedale, Ill., breeders who in 1882-3 and -4 were showing Durocs at the principal western shows.

J. M. Browning's foundation herd was eleven sows, purchased from A. Ingram, being a combination of Ingram's and Bradbury's breeding, the latter a resident of Virginia. One of these eleven sows was later the dam of British Wonder 917, a famous show hog in the Browning & Son herd, grand champion at Illinois fair, 1890, and other early shows. British Wonder 917 was by Perry Duke 185, the latter coming from the Tom Bennett herd and was sired by Red Chief, unrecorded, who was also the sire of Red Jacket No. 1, the first Duroc boar ever put on record. The dam of British Wonder was Queen of the West 530, a sow from the Ingram herd.

William Roberts' first recorded herd boar was Alexis, from the same herd mentioned above (Railsback & Pittsford), and was first at St. Louis, 1882, second at Iowa, 1883, Nebraska 1883. Later he used Herman H. from the Herring & Hummer herd, a boar tracing to Climax 21, winner at New Orleans in 1885 and seven big fairs in 1884; further on in his work of constructing Durocs, Mr. Roberts used Delay 3167, by Troubadour and Supply 4563, both of which he bought from the Morton herd, the latter being sired by Col. M., the foundation of the Colonel family. Columbian Duke 3457 was also raised in this herd in the early 90s and his get was at Chicago, his breeding tracing to the C. J. Stuckey herd of Illinois (now a breeder of Ohio).

Mr. Roberts also used Exchanger 2539a-159n, bred by N. Harrington of Iowa, and sired by Imperial 2nd, by Monas Imperial and out of the sow Mona 2nd. The dam of Exchanger was a full sister of his sire, making Exchanger an inbred boar. This boar was later sold to Mr. Stonebraker of Illinois and was shown in Chicago in 1893, winning in the aged class, and he also placed several of his get in the winnings. Exchanger was an inbred Holmes boar, tracing through both his sire and dam to Mona 2nd, winner at New Orleans in 1885, Des Moines '83, and Ottumwa '84, and Chicago at the old exposition in '81. Further along in the Roberts' herd Allison was purchased of Walter Abernathy of Indiana and about the time of the World's Fair in Chicago Mr. Roberts purchased the great show sow Ohio Anna 10068, foundation sow of the Ohio Anna family and dam of the Orions. Roberts & Sons' name became synonymous with good Durocs, and this firm was a leader in Iowa improvement for a good many years. Old Orion was purchased by them in 1895 and crossed with Ohio Anna 8th, producing Orion 2nd.

The brightest light that guided the way for Duroc improvement in pioneer days and the star that led through many trying years, a man to whom Durocs and Duroc breeders owe more than to any one man for breed building and for swine leadership in his day and time, came upon the scene of action early in the '80s and by his knowledge of mating and his master mind in constructive breeding, laid the foundation for several of the most noted families the breed has ever known. The world has produced no greater constructive swine breeder than Sam Morton of Ohio. The foundation of his recorded herd was Derby 3962, Molly Ann 3964, Mora 3966, Java 3968 and Vernal 2nd 3970, purchased from the Browning herd in Illinois, and on the same trip to the Sucker State he purchased Champion Wonder 1299 from Andrew Ingram. The sows mentioned were all daughters of British Wonder 917 and out of Vernal 998, tracing to Nero 205 from the William Holmes herd. Champion Wonder was by the same sire as British Wonder and out of Success 2nd 2668, by Perry Duke, the latter sire of both British Wonder and Champion Wonder. The famous Duchess sows of the Morton herd were founded on one of these sows, Vernal 2nd 3970. She was mated to Granville 2nd

1301, whose dam was by Perry Duke, and who on his sire's side was a great-grandson of Perry Duke. Of this cross came Vernal 2nd's daughter. In four generations of Vernal 2nd's daughter's pedigree, Perry Duke, the foundation boar, appears four times. Vernal 2nd's daughter was mated to King Stoner 1661, producing Duchess 2nd 5932. King Stoner was from the G. W. Stoner herd of Illinois, was sired by Prince Apollo, a boar of the Holmes and Clark pedigree. Duchess 2nd was later mated to Troubadour 2255 by Doak 2nd, bred by Mr. Doak, of Iowa, and produced Duchess 9th. Duchess 9th was second aged sow at the Columbian Show of '93. There is little doubt that this intense breeding brought about by the commingling of these blood lines and by the adding of an occasional outcross produced as good a family of sows as has ever been known. Duchess 9th was bred to Colongues by Legal Tender, by King Stoner; Legal Tender's dam being by Morton's Perfection, a sow tracing directly back to Mora, one of Morton's foundation sows. Duchess 19th was produced by this cross to Colongues, and she was mated to Storm King, a boar tracing directly back to the Duchess sows and Duchess 32nd resulted from this cross. Duchess 40th was produced by the mating of Walts Col. to Duchess 32nd 13364, just mentioned in the preceding sentence. This latter cross threw in another tracing to Troubadour, another to King Stoner, and another still to Mora, the show sow mentioned previously. Duchess 40th 18958 was mated to Protection and produced Ohio Chief and Chief of Ohio. She was mated to Orion 2nd and produced Top Notcher, the head of the Top Notcher family, and was mated to Morton's King and produced Winchester Chief 10077, a boar that added size and strength to Indiana Durocs.

Thos. Bennett of Illinois bought few boars from other breeders, preferring to line and interbreed, for during the '80s and early '90s we find but two boars in use in his herd from outside breeding. One was Oswego 665 by old Dan Voorhees, unrecorded, mentioned heretofore as being used in the Railsback & Pittsford herd. The other was Legal Tender 2179, purchased from S. E. Morton after the latter breeder had used him some time. Boars of his own breeding used in these dozen or more years were Chickasaw Chief 1551, Choctaw Chief 1553, Conqueror 2181, Joe the Banker 1955, Cuckoo 3063, and the latter's son, Hero 1955, grand champion boar in his under year form at Chicago, '93. The get of Cuckoo showed splendidly at this fair and the old boar himself lacked fitting or would have given Col. M. a hard tussle for first honors in the top class. Choctaw Chief was out of Minnesota 702, a winner at the old exposition in Chicago in '81.

Amos Harris of Kentucky was the most southerly located of any breeder recording in these early days, and he was using Lafoon 1315 from the Stoner breeding; Morton 2305, a King Stoner boar, from the Morton herd; Edgemont 2625, a Troubadour boar from the same herd; Hummer 3371 from the Herring & Hummer herd, and Guy Wilkes 3459 from the Ingram herd.

G. W. Stoner of Illinois used old Breckenridge 387, a boar produced at Breckenridge, Mo., and a boar recognized at that time as having wonderful stretch and size and vigor. Prince Apollo, his son, was also used in this herd and later on Royal Duke 1357 from Rankin & Son of Illinois; Camden Prince 1615 from the Morton herd, sired by Champion Wonder 1299, brother to British Wonder, was also used.

Mr. J. H. Lathrop of Iowa used in his herd a boar by Iowa Champion, from the Hubbard herd, and sows from the Stuckey and Morton herd.

Our object in giving these paragraphs regarding these boars, is to bring to your mind the amount of buying and selling of herd boars that was going on between the different sections of the country. With the few exceptions we have given you the breeding and the boars used in these herds for ten or twelve years before the Columbian Exposition, 1893, the first place that the Durocs had been accorded a place worthy of their standing in any show of world wide significance. The show at Chicago brought together the herds of Ohio, Illinois, Nebraska and Iowa. The show was judged by J. A. Countryman, of Rochelle, Illinois, who was assisted by D. J. Spaulding, of Black River Falls, Wisconsin. The herds represented were that of Morton, Bennett, Searle, Stonebraker, Chandler & Son (Nebraska), Walter & Bro., Roberts & Son, G. W. Trone, J. H. Lathrop. There was a variety of type found among the boars of the several classes but from the

best account of the show put in print at
that time, for which we are indebted to
the Breeders' Gazette, it would seem that
the judges succeeded in placing animals
worthy of the position he gave them.
Hero, the Grand Champion boar, was
surely best of the male section, and was
awarded that place after being placed at
the head of the under-year class. The
sows of the show, as they do many times
in our later contests, made more pleasing
appearance and showed more even type.
Lucy Wonder, the head of the great Wonder family of sows was placed at the top
in the aged class. Duchess 9th, mentioned heretofore in these pages, being
second. Some mighty choice things in
the pig class came to light from the Ben-
net herd, daughters of Cuckoo, already
mentioned. They showed considerable
length, well backed, and hammed and lots
of finish and quality. In closing the account of this sow the Breeders' Gazette
states that the judge would not have gone
wrong if he had turned Lucy Wonder
down for Champion and placed the purple
on Bennett's under-twelve-month pig,
Brightness.

Two years after the Columbian Exposition old Orion was farrowed on the N. P. Clark farm at Monticello, Iowa, and was shown with his litter mates at Des Moines that Fall. The history of this boar and his litter mates and their progeny form a prominent part of Duroc lore from 1895 to 1900.

About 1890 in Northwest Iowa another breeder of Durocs was gathering a herd destined to play an important part in developing the good feeding qualities and increasing the popularity of Durocs in the Corn Belt. O. S. West, of Iowa, gathered his breeding hogs from the various good herds. Among his first boars was one called Crimson Wonder 2813, by Trojan, from the Roberts herd. He also used Banker Boy from the Morton herd, he by Geo. W. S., by old Breckenridge, heretofore mentioned; Banker and Highland King, by Hoosier King, from the Stuckey herd. Czar Nicholas, by Coulongues. Some excellent constructive breeding was done in this herd.

In the latter '90s many boars were changing owners back and forth from East to West and vice versa. Showing at state ·fairs had continued with the impetus given by the Columbian Exposition. At one

time Morton & Co., a firm composed of
S. E. Morton, Walter Abernathy and E.
M. Borradaile, bought four boars from
Wm. Roberts & Son. Two of these were
Allison, Jr., and Hustler, sired by Allison
5267, the latter sired by Panic 4107, a boar
that Roberts had purchased from Morton
& Co. a few years before. They also
bought Orion II 6537, by old Orion, out
of Ohio Anna 10068. Another one was
Bally Trally 6545, also by Panic. By this
purchase they secured boars of practically
the same line of breeding of their sows
and intensified that blood in their offspring.
This firm had previously produced, namely
in '94 and '95, Walts Col. 5795, and Protection 4697, the former a son of Col. M.,
and the latter a son of Coulongues, the
Bennett herd header. Around the story
and names in this paragraph lie the foundation of the Protections, the Colonels,
and the Orion Chief families.

Besides using old Orion and Allison at
this time the Roberts were using to good
advantage the hog called Awake, by John,
the latter a son of Cuckoo 3063, from the
Bennett herd. Orion, Jr., about this time
went to an Indiana herd as did Orion R.,
both out of old Orion. These went to

Finch and Pearson, of Indiana. Roberts also bought a Protection boar called I See. They also used Zeek, a winner in the pig classes at the Columbian in '93. J. W. Stribling, of Iowa, had a Protection bred boar called Ohio Tom, as did also J. Benson & Son, of Iowa. During these years from '80 to '90, the Browning herd in Illinois was using such boars as Col. Wonder 3817, by British Wonder, and a grandson of Iowa Champion 2443, and General Lyons from the Morton herd. It was during these years that the son, Harry E., so well known in later years in the Duroc world took the reins of the herd and continued to build one of the most famous the breed has ever known. Columbian Chief, by Col. Wonder, by British Wonder, Vermillion Prince, by Col. Champion, and Walters Onward from the Walters herd, were used.

About this time, Walter & Bro., of Ohio, had developed Tom Wonder 6061 and U. S. Wonder 6057, out of Lucy Wonder 6334. These breeders perfected the Lucy Wonder family and the Walts Duchess family of sows. Mr. Mahan, father of the present Duroc man, C. E., founded his wonderful herd of sows on a litter sister

of Lucy Wonder. In the Bennett herd,
Referee, by the Grand Champion Hero,
was being used to good advantage and the
blood of this Grand Champion was dif-
fused into many of the prominent herds
of the time, one son going to the Reed
herd in Iowa in '99. In Nebraska the
Searles had developed Aksarben, Jr., and
in Ohio the Mortons had perfected the
Col. M's Variety family of sows and had
added a boar from the Stoner herd called
Royal LeGrande, a boar tracing back
through the H. W. Mumford herd of
Michigan and a boar that was almost an
entire outcross for Protection and Col. M.
sows.

It was late in the '90s when another man
entered the Duroc arena, a man who was
destined to be a leader, a man who had
the instinct of the constructive breeder and
the knowledge of the requirements of a
good hog and had the tenacity, vigor, and
push to get rapidly towards the front and
during all the years from the time he embarked in Duroc work and got well under
way, his herd has been in the lime light
and many noted breeding animals
have gone out from this headquarters.
The first recorded animals owned by Ira

Jackson, of Ohio, was Longfellow 6815, from the Walter & Bro. herd and Lord Clinton, a Col. M. boar, from the Morton herd. He also had sows from the C. C. Brawley herd and the Morton herd, and it was one of these latter, Agnes X. 15250, that he crossed with old Longfellow that produced Mabel, the dam of Orion Chief, head of the Orion Chiefs and the Orion Cherry King families, so popular at this writing. The sire of Orion Chief, Orion II 6537, and Chief of Ohio 9775, a litter mate to Ohio Chief, were purchased from Morton. Surprise, probably one of the greatest individuals produced in the early part of 1900, was by Chief of Ohio, out of Mayflower, a daughter of Lord Clinton.

One of the early boars of the R. C. Watt herd in Ohio was Moquette 3415, a son of Duke of Star Herd, a winner at the Columbian Exposition. Mr. Watt had purchased him from Walter & Bro., as he had also Dora Wonder, a daughter of Lucy Keever, and in this herd King Watt was also used. Bob Watt developed the sow family of Cedarville Queens, the original Cedarville Queen being a daughter of Van-Ausdal's Best, a sow tracing to Hoosier King through the Abernathy herd and the

Tom Bennett herd. The sire being Cannon Ball, tracing directly back to old North Star in the Stoner herd in Illinois. Uncle Tom, a boar from the Tom Bennett herd, was next used, he by Chicamauga by Rattler, a boar from the Stoner herd. From the Morton herd Mr. Watt purchased Top Notcher, then from the cross of Top Notcher and the Cedarville sows, many noted boars were sent to different sections of the country. One of these that placed considerable improvement in the Western and Northwestern Durocs was Malcolm Model, a boar going to Wm. Malcolm, of Minnesota.

It was about this time that Commodore 13385 came upon the scene in the Bennett herd. He was produced practically from the Bennett breeding, from early foundation animals, and aside from Hero, the Grand Champion at the World's Fair at Chicago, 1893, was the most prominent boar ever in the Bennett herd.

During this period of the eight or ten years which we have covered in reciting the advancement and improvement in the blood lines of the herds in Ohio and other Eastern sections, the interest in the West had increased by leaps and bounds and hundreds of new men were laying the foundations of herds which produced the animals whose history leads so intimately into the story of Durocs in Western territory. It was also during this period that the public sale idea of pure bred animals was developed in the Duroc breed, and as far as the writer is able to learn, this was developed both in the West and in the East at about the same time. One of the first public sales attempted was at the Iowa State Fair grounds in 1894, by J. H. Lathrop, of Oxford Junction. He was his own auctioneer, and compelled to stop before the sale was over on account of a lack of patronage. In the winter of '96 this same man had a closing out sale, the average being about $35 or $40, the top sow bringing $60, and purchased by O. S. West, of Paullina, Iowa. Mr. Wm. Roberts & Son had held some early sales along about the same years between '96 and '98. I think the first public sale of Durocs in the central hog states, at least in Illinois, was the same of Geo. W. Trone in 1896.

In 1898, Mr. J. C. Woodburn, of Maryville, Mo., held his first sale. In the East, Morton & Co. lead with these public vendues, and Dr. Burkhardt, of Cincinnati,

Ohio, held several sales as did several combinations of Ohio and Indiana breeders
later on. It was not until after the World's
Fair at St. Louis in 1904 that these sales
throughout the hog belt became general
and were adopted by nearly all of the leading breeders.

During the interim between the Chicago
Exposition and the St. Louis Exposition
a world of new herds were developed. Es-
pecially was the business active in the
West in the territory where more hogs are
raised to the square mile than any place
else in the world. Such men as Geo.
Briggs, Gilbert VanPatten, H. B. Louden,
Smith and Arch Brown, of Nebraska; R.
J. Harding, Johnson Bros., H. C. Sheldon,
John Henderson, of Iowa; McFarland
Bros., W. L. Addy, J. C. Woodburn, of
Missouri; John L. Hunt, of Marysville,
Kansas, and many others had come to the
front with good herds, among them being
G. W. Seckman, of Illinois, who attended
his first public sale at Geo. W. Trones'
and paid $85 for a sow called Grace Dar-
ling, and laid the foundation of a great
herd. He later attended the Andrew In-
gram sale and bought the sow, Hattie Gold
Dust, that laid the foundation for the Gold

Dust herd and family of sows which were so well represented in the winnings at St. Louis in 1904. A. F. Russell, of Missouri; C. W. and Wm. Reed, Iowa; S. Y. Thornton, Missouri; C. R. Doty, Illinois; O. W. Browning, O. N. Woody, Harry Pfander, E. P. Watson, O. E. Osborn, S. E. McCullough, David Nauman, A. P. Alsin, H. L. Cantine, Iowa; T. L. Livingston, Jacob Wernsman, the Manleys and Wm. Stufft, of Nebraska; Kraschel & Son, of Illinois. All these were building strong herd foundations as these years passed.

Two of the early boars of Nebraska, to which the breed is indebted for improvement and especial recognition by the men who were growing pork hogs, were Higgins' Model 3251, and Improver II 13365, both used in the Geo. Briggs herd, and their blood was widely scattered through the West. Higgins' Model came from a combination of Indiana and Iowa breeding. Improver II traces remotely to Coulongues, the sire of Protection, the head of the Protection family, and his dam traces through the Lathrop herd in Iowa to the pioneer herd of Bennett, of Illinois. Both these sires lived to a ripe old age, and their progeny always found ready sale.

Other boars used by Briggs in an early
day was Briggs' Choice, a descendant of
Trones' Hero (winner, Chicago, '93), Red
Chief I Am, a son of Protection, Billie K.,
from the O. S. West herd; Kantbebeat,
Lincoln Wonder, by Ohio Chief; Billie's
Wonder, Briggs' Crimson by Belle's Crimson Wonder; Crimson Wonder 4th, by
Crimson Wonder Again. He also used
a Golden Model boar by Golden Model
Again. The Briggs herd has for twenty
or more years been a leader in Nebraska
in producing improvement in Durocs. During the past few years, Illustrator II by
L. E.'s Valley King, and Joe Orion 5th by
Joe Orion 2nd, and a Great Wonder boar
have been in use.

With but few exceptions the main herd
material we find recorded in the National
record at that time in the West came from
the herd boars whose blood lines run back
to the Roberts, Clark and Hubbard herds of
Iowa, tracing to boars founded on the best
blood they could buy and produce any-
where in the hog belt. Old Crimson Wonder had been produced on the Cantine
Farm in Iowa and sold to A. T. Cole, of
Nebraska, and in the latter's closing out
sale had gone to W. A. Kirkpatrick, of

Lincoln. This boar is head of the great Crimson Wonder family that cut more figure in the breeding of the Western herds for the next twelve years than any other one blood line. He changed hands in the closing out sale mentioned above at $695, the highest price at that time that any one ever dreamed would be paid for a Duroc boar.

Just before the World's Fair in St. Louis one of the men in the West destined to cut more figure in the production of good Durocs and to lead in more shows for a longer period of years than any one of the Western breeders, and who is still actively engaged in the business, tossed his hat into the ring and has ever since that time been a leader in not only shows and sales, but in real constructive breeding. This man is J. D. Waltemeyer. In Volume 13, (not unlucky in this case), you will find the first boar recorded in his name, No. 21523, Iowa Chief. This boar was bought from the Woody Herd, and he was a boar ahead of his time. His sire came from the Addy herd, tracing back directly to that great sire, Awake, of the Roberts herd, one of the foundation boars of that herd. Waltemeyer later used a boar from the Russell herd in Missouri, Bobby, a descendant of old Referee, by Hero, the latter Grand Champion at Chicago, 1893. Also a Malcolm Model boar and a boar called Bugal Boy, by Keep On IV, tracing to Reed's Banker, a boar that did excellent improvement work in Iowa Durocs, a boar sired by old Advance, sire of Proud Advance, and whose ancestry traced through Tacoma to Old Orion. Model Boy, a Malcolm Model boar, was also used. In 1908 he bought Golden Model 2nd and since that time the Waltemeyers have builded one of the best strains the Durocs have ever known—the Golden Models, using sows of Ohio Chief and Crimson Wonder breeding with a dash of Colonel blood and such boars as Model Chief's Last, by Model Chief 2nd by Model Chief by Ohio Chief; Sensation Wonder, grandson of Crimson Wonder Again, and Golden Model 31st, I Am Golden Model 2nd, Grand Model and others. More recently they brought into the herd Great Wonder, a boar bearing the blood of Ohio Chief, Crimson Wonder, Orion, the Colonels and Critics, a boar of splendid feeding and breeding qualities. The Golden Models through the Waltemeyer efforts took a lead in Western Durocs, and are admired for their smoothness, symmetry, and wonderful feeding qualities. Among the show and breeding boars produced from Golden Model II are Grand Model, Golden Model 11th, Golden Model 34th, Chief Model 2nd, High Model, Blue Ribbon Model, Chief Model, Golden Model 17th, Chief's Model III, Golden Model 31st, Foxey Model, Golden Model 30th, W. B.'s Golden Model and Golden Model 5th. I Am Golden Model 2nd was one of the good breeding sons of Golden Model 2nd.

The St. Louis Duroc Show, 1904, seems now like a small show in numbers with only 348 head on the grounds. A larger number of the breed had been counted in several State Fairs in the West prior to that time, but it brought the best from nearly every section of the country, and it was a really high class show, for the time which the breeders had been constructively engaged in perfecting this new breed of hogs. It was the first time that the representative farmers from all corners of the globe had the opportunity to see real Duroc hogs in their good form, and I have heard leading men of other breeds say that our show at St. Louis set the pace for the large type hog which is in such demand today, and for which every one is striving. Our breeders at that show exhibited strong, vigorous, somewhat coarse, boars, that showed that they were real producers and they attracted the attention of the visiting farmers and feeders. The names of the winners in that show and the names of their progeny have become household words wherever Durocs are known. Such hogs as Ohio Chief, Kruger, Top Notcher King, Kant Be Beat, Tip Top Notcher, Commodore, Jumbo Perfection, Orion Chief, High Chief, Crimson Wonder, Advancer, Top Notcher Chief, Medoc, Model Chief, Tom Watson, Dotie (Grand Champion sow), Xenia Belle, Roberta, Lady Advance, Lady Orion, Daisy Improver, Cedarvale Queen 8th, Vic's Bell, Crimson Lady, The Genevieves, Cols. Belle 2nd, and S. E.'s Model 2nd. These are only a few of the more than 300 winners in this show whose places were picked by J. A. Shade, of Kingsley, Iowa. There was $5,110 hung up in the regular classes by the World's Fair. The National Association gave $1,000 more, the American a silver trophy, and there were any number of prizes furnished by the State Legislature.

Johnson Bros., of Newkirk, Iowa, won the Champion pen of barrows. The contenders for that purple show being H. S. Allen, of Iowa, with a pen of six and under twelve; C. R. Doty, of Illinois, on a pen under six. The champion barrow went to C. R. Doty. The champion barrow, any age, went to Doty. Messrs. Trone, of Illinois; Sheldon, of Iowa, being the winners in the other classes contending for the championship. The champion grade by recorded sire was won by Trone, of Illinois. There were twenty-one members of the National who participated in the special prizes hung up by that organization. The following are winners of 1st and 2nd prizes and Grand Champions in the show:

Aged Boar—13 shown; 1st, Ohio Chief, owned by S. E. Morton & Co.; 2nd, Kruger, owned by Geo. W. Trone. Top Notcher King, Kant Be Beat, Top Notcher Again, Billie K., King of Kings, The Lad For Me and others were lower in the class.

Senior Boars—12 shown; 1st, Tip Top Notcher, owned by Geo. Seckman; 2nd, Commodore, owned by Thos. Bennett. Among the others in this class were Jumbo Perfection, Kant Be Beat Again, Gay Advance, Searles Olympus.

Junior Boars—18 shown; 1st, Orion Chief, owned by Ira Jackson; 2nd, Goldie's Top Notcher, owned by Seckman. Among the others shown were Chief Orion, High Chief, Belle's Chief, Matchless Chief.

Senior Boar Pigs—23 shown; 1st, Joe, owned by McFarland Bros.; 2nd, Perfection Chief, owned by Gilbert VanPatten.

Junior Boar Pigs—32 shown; 1st, Top Notcher Chief, owned by J. A. Teter; 2nd, Checkmate, owned by Seckman. Medoc, Tom Watson, Lord Gold Finch, Ideal Top Notcher, Ohio Col., He's Our Pride, afterwards well known, were shown in this class.

Aged Sows—19 shown; 1st, Dotie, owned by McFarland Bros.; 2nd. Walt's Belle, Morton & Co. Nellie A., Brooklyn Mabel, Daisy Maid, Rubertha, Walt's Model and Bessie H., were in this class.

Senior Sows—14 shown; 1st, My Choice, Ira Jackson; 2nd, Moss Rose, McFarland Bros.

Junior Sows—19 shown; 1st, Lady Orion, Ira Jackson; 2nd, Chic's Belle, Morton & Co. Daisy Improver, Nellie Wonder II, Orion Lady and May Advance were sows that became great producers also in this class.

Senior Sow Pigs—30 shown; 1st, Kant's Best, Watt & Foust; 2nd, Cedarville Queen VIII, same. Orion's Choice, S. E. Model IV., Lady Banker and others in this class came into prominence later.

Junior Sow Pigs—37 shown; 1st, Crimson Lady, by J. E. Mendenhall & Sons; 2nd, Top Notcher's Best, J. A. Teter. Our Pride II, Queen Genevieve, Fashion Queen XII, Cols. Belle II., S. E. Model VII., and others in this class are well known to modern Duroc men. Our Pride II., in the Watt & Foust herd, and later in the McKee herd, produced excellent get.

Grand Champion Boar — Tip Top Notcher, Reserve Champion, Top Notcher Chief.

Grand Champion Sow—Dotie, Reserve Champion, Kant's Best.

A perusal of these winnings will show that the Top Notchers had the best of the winnings. Old Top Notcher's descendants were represented in seven of the first and second prize winners: Tip Top Notcher, Goldie's Top Notcher, Top Notcher Chief, Checkmate (by Tip Top Notcher), Kant's Best (dam by Top Notcher), Cedarville Queen VIII (sired by Top Notcher), and Top Notcher's Best, and his progeny held Grand Champion and Reserve Champion boar. Top Notcher 8803 was sired by Orion II., out of Duchess 40th; Orion II., by Old Orion. By adding the Jackson winnings to the above the descendants of Old Orion led by quite a margin.

That the public sale had come quickly and had come to stay was evident when we find that something like 75 auctions of Durocs were booked for the winter of 1904-05, immediately after the World's Fair. Several Fall sales had been held previous to that. Notably among these was that of J. D. Nidlinger, of Decatur, Indiana, on Nevember 1st, where the prices ranged from $15 to $135, the top price being paid by Smith Brown, of Nebraska. History was written very rapidly during the next few months. One of the noted sales of that winter was sows bred to Proud Advance, averaging $80. Proud Advance was a combination of Orion and Protection breeding. His sire, Advance, being by Tacoma, he by Orion Chief 6601 by Old Orion and out of Ohio Anna 10068. Proud Advance's dam was by Malcolm Chief 7811 by Jumbo Red by Protection and his second dam traces to Legal Tender, a foundation boar used by Morton and Stoner. The breed has never seen the superior of Proud Advance as a sow sire, and this wonderful stretch and smoothness had much to do with advancing the popularity of the breed. He was bred by Johnson Bros. and Reed, of Iowa, was later owned by H. C. Sheldon and the Manleys. The highest priced sow of the breed, and many others that sold well up to the record top, carried his blood.

Kruger, winner of the second prize in the aged herd at St. Louis, was purchased by Jackson in the Trone sale for $550 in February, 1905, and his get were widely scattered, Iowa breeders securing one or two of his noted sons. Nebraska Belle sold that winter in the Brown sale (Nebraska) to R. J. Harding for $600, and the next day he refused a check for $1,000 for her from Gilbert VanPatten, who had started for the sale to buy the sow, but was blocked by a blizzard.

Old Orion made his owners, Manley & Co., an average of $73 the same season. In Ohio the breeders who had shown at St. Louis staged a two-days public sale and one sow of Cedarville Queen breeding sold for $350. Others sold well up above $100.

J. D. Nidlinger, of Indiana, had won first at Ohio and Indiana with Hanley, an under-year boar by None Such, from the Stoner herd, his dam being from Echo King breeding, and sold him at public auction for $310 to McNeil Bros., of Illinois. Echo King blood was later diffused in Nebraska and became quite popular.

Crimson Wonder I Am, a son of old Crimson Wonder 26355n, was Grand Champion, 1905, at Iowa, thus continuing the champion record of the family begun by his sire two years before.

Buddy K. IV., was Grand Champion of Illinois, 1905, one of the largest, coarsest, ruggedest sires the breed has ever known, and while he was lacking in quality he had many characteristics that the breed needs —length, height, big bone and an all-over big frame. After being used more than a year at the head of the Ed Baxter herd, after winning his Championship, he was sold on Feb. 6, 1907, for $5,025 to McNeil Bros., and today still holds the record of being the highest priced boar of the breed sold at auction.

In the winter of '05 and '06, in the

Browning sale, Helen Blazes III sold for $1,000, the highest priced sow at that time. She came from a producing family of sows and her new owner afterwards sold one of her gilts at private sale for $1,200. Browning's sale, in which this sow sold, averaged $110.50 on 39 head. Eastern sales were increasing in interest and such prices as $121.50, paid by Col. Igleheart in the Morton sale, and $317.50 by Thos. Johnson in the Whitehall sale, were not uncommon. In the Henry Allen sale in Iowa, Allen's Maid brought $300, and R. J. Harding bought Proud Lady by Proud Advance for $1,275, bred to Old Orion in the Manley sale.

The leading Fall shows of 1906 produced such champions as Oom Paul in Illinois, a boar owned by George Trone and sired by Jumbo Red, he by Protection. Jumbo Red was a noted son of Protection that went into Nebraska and did much to popularize Durocs by producing not once, but several times the top car lots on the Omaha market. Crimson Wonder Again was champion at Iowa, making three generations of grand champions early in the history of that noted strain.

In Ohio King's Pal, by King to Be, by

Morton's King, won the purple, and was sold to Thos. Johnson at $1,000. He lived only a few months, but left some unusually good get, among them King Pal's Last, a boar that headed the M. D. Harvey herd in Indiana, and sired Pal's Prince and Colonade, Indiana Grand Champions; Colonade being also Champion at the Chicago International, 1912. At Illinois, Commodore produced from the Bennett herd, but purchased by Harris & Son, of Kentucky, won the purple and was at once popular with the breeders on account of his size, vigor, bone and quality. He had come to Illinois after winning at Kentucky and Tennessee. In the winter following, his owners made an average of $179.50 on sows bred to him. Although he produced some good sons and daughters, he was a disappointment as a real herd header.

January 12, 1906, Old Orion died in Nebraska at the age of eleven years an undefeated boar. He was shown as a pig in 1895, as a yearling in '96, and was Champion at Iowa in '97, and a winner at Omaha, 1898. Roberts & Son sold him to S. E. McCullough, of Iowa, who later sold him to Manley & Son, on whose farm he died. The Orion Cherry Kings and many of the Crimson Wonders of today trace directly to this great sire.

In November, 1906, a deal was consummated that gave a decided impetus to Duroc enthusiasm, both East and West. R. J. Harding and O. E. Osborn, of Iowa, purchased of S. E. Morton & Co., Ohio Chief, a boar sired by Protection, and out of the noted sow, Duchess 40th, mentioned in earlier pages of this history. Ohio Chief won as a pig and as a yearling at the Chicago Exposition in 1900 and 1901 and as aged boar stood at the head of his class at St. Louis. He undoubtedly was the best boar the breed had produced, as well as the biggest, up to this time, and although Tip Top Notcher was given the purple at St. Louis on account of his showing more excellent flesh, many at the ringside had picked the aged boar for the purple. In Harding's first sale after he purchased the boar, sows averaged $98.30 with a $730 top. Later the owners sold a third interest to J. M. Morrison, of Nebraska, for $2,000, the same amount they had paid Morton & Co. for the hog. Morrison broke a partnership agreement by selling 18 sows bred to him in one sale and Messrs.

Harding and Osborn bought him out and Ohio Chief died on the Harding farm.

Tip Top Notcher, the World's Grand Champion at St. Louis, produced many good sons and daughters, Helen's Tip Top Notcher, one of the former, selling at a record price as a yearling at $1,000. He transmitted to his progeny much ruggedness and prolificacy. Geo. Seckman, who owned and showed him, sold him for $5,000 to a company of Illinois breeders, but before the hog died, Seckman re-purchased him. Tip Top Notcher was by Top Notcher 8803, he by Orion II., by Old Orion, and was bred by R. C. Watt, of Ohio. His sire was a winner in Ohio shows, and his dam came from the Morton line of Variety sows. A controversy over the correctness of his pedigree came up immediately after his winning at St. Louis, and the American Association at one time voted to not accept pedigrees from this breeding, but the following year rescinded their action, the directors not considering the charge backed up by sufficient evidence to warrant the former action. The officers of the National Association found insufficient grounds for any action, and its Board accepted all pedigrees of his progeny. Tip Top Notcher is one of the few Durocs who boasts a marble slab to mark his resting place. Close to the main road on the Seckman farm in Brown County, Ills., the headstone is plainly visible.

During the winter sale season of 1906-7 that inflation of prices that seems to come in cycles, and which unerringly causes disaster to breeds of live stock wherever it attacks, took possession of all quarters of the Duroc selling and laid the foundation for the panic and depression in which many breeders were mired. The bank panic coming late that year, augmented the losses and depression. No amount of warning, nor numbers of wrecks visible along the route of the pure bred industry seems to be of any service in such times, and although the breed paper threw out signals against these "frameups" and fictitious prices, and combinations invented to boost prices, little or no heed was given. We believe space given here to one of the breed paper's articles will be well worth the space taken, for it applies today as it did in 1907. This is the article:

"Why readers and breeders should get a wrong impression from plain English we don't know, but it seems some have.

Our stand on the sale business is this: We stand ready to encourage every breeder who desires to make a sale of good, well bred animals, we don't care how high the prices are, so the money or its equivalent changes hands. We are strictly against 'frameups' no matter whether made at a big sale or a beginners' sale. No price is to high for an outstanding individual with blood lines that have shown worth. No price is too low for a mean, inferior animal, no matter what the blood lines. 'Suckers' are not always men who pay the high prices, for the men in the business who have made money are the ones who have paid high prices for good animals. 'Suckers,' so-called, are men who have been influenced against their better judgment to pay more than they can afford to if the animal should die. If there has been anything published by us that can be interpreted as a discouragement to the young breeder, please cite us to the article. We are for high prices, but not 'balloon' prices. Somebody will stick a pin in these prices some day, and then what?"

Rosebud Lady sold for $1,575 in the Briggs sale, an open Proud Advance sow brought $500 in the Browning sale. Alix II, the Iowa and Nebraska Champion, sold for $2,200 in the Morrison sale. One-half interest in Kant Be Beat sold in the same sale for $1,500. Alix II was a sow bred by Manley & Co., sired by Proud Advance out of an Orion dam. She never proved to be a real producer. Kant Be Beat was by King Hustler out of a dam by Longfellow 6815, and was bred and shown by Watt & Foust. He had forty or fifty Championships to his credit., and was a good boar for his time. Ceres Belle, a noted Nebraska sow, brought $1,025. S. E.'s Model 30996, by Ohio Chief, a sow shown at St. Louis and sold to the Proud Advance Stock Co., was sold in their sale at $1,060.

Red Wonder, a Pilot Wonder bred hog, Owned in Illinois, was Grand Champion at Iowa, 1907, and Ruberta G., full sister to Rubertha, the 1906 Champion, won the female purple. Red Wonder was not an even producer, but sired a few choice daughters. In Indiana, Col. Scott by Carls Col. was Grand Champion, and M. A.'s Model, shown by the Kraschels, of Illinois, was sow Champion. She afterwards sold in their sale for $1,001. She was a line bred Ohio Chief sow. At Nebraska,

Lincoln Top, shown by Putman, won the lead, and Clay Center Belle, owned by Briggs, was at head of the sow classes. Lincoln Top traced to Red Chief I Am, a Protection bred boar. Kelly's Pilot Wonder won the purple at Ohio, and went to Whitehall Farm where he proved to be a great producer. He was a grandson of Old Pilot Wonder, a Jackson show boar. Pilot Wonder was by U. S. Duroc by Walt's Col. out of a Lucy Wonder sow. Inventor, by Climax II, won at Illinois, 1907. He was a McFarland bred hog, coming from their herd to Illinois as a pig. He was sired by II Climax, sire of many of the McFarland winners. Gatton's Perfection was Grand Champion sow at Illinois, 1907, and afterwards sold at auction for $1,200, a Kant Be Beat boar selling in the same sale for $1,295.

During the winter of 1907-'08 the prices tended to rise and among the notable sales were Savannah Belle and her litter of Ohio Chief pigs at the Sam Murphy sale for $3,300, he having paid $780 for her at the Harding sale. She was a daughter of the noted Nebraska Belle. Ruberta, dam of the Kant Be Beat boar. Dreadnaught, sold to Manley & Co., for $2,500. Proud Zeda by Proud Advance, sold in the Browning sale, breaking all sale records, at $2,600. A half interest in the Champion Lincoln Top went for $2,650 in the Putman sale. Sexsmith & Strong sold a granddaughter of Nebraska Belle for $2,100. A sow was sold in S. E. Morton's sale for $2,800. An Eds Col. sow in the Jackson sale brought $1,000. Johnson Bros.' sale in Iowa had a $750 top. Some of these animals made good and their progeny is still going on record, but many were sold under enthusiasm and excitement and had little to recommend them as valuable additions to herds.

King of Cols. 16075 was ascending the scale of popularity about this time, and a complete history of his progeny for a period of several years would cover a goodly portion of the breed's history for that time. He was bred by Carl Scott, an Indiana breeder who did much to propagate good Durocs and incidentally promote that leading family, the Colonels. Mr. Scott owned Eds Col., and from him produced Carls Col. and in turn Prince of Cols. In the Fall of 1905 he showed at several fairs in Indiana a litter of Prince of Col. pigs out of Love 35060, among which was the boar afterwards known as King of Cols. He was never better than third, Muncie Chief and others of the litter always leading him. When Scott had a sale in the Fall, Muncie Chief went to Colbert & Stroud, of Indiana, and was the first pick of a majority of the breeders present. The old "War Horse," Sam Morton, attended the sale and had picked out this undeveloped pig as one worth while, and when he bid him in, a good many who thought they knew hogs, wondered at his choice. This hog brought Morton more notice and more money than any he had ever owned, and breeders from all quarters of the hog kingdom were soon coming to him for King of Col. get. A syndicate of Illinois breeders offered him $10,000 at one time for the boar, but the offer was turned down, and the boar died on the Morton Farm in the height of his usefulness. Among the show boars he produced were Chief's Col., Defender, King of Cols. Ideal, Col. S., King Wonder, Ohio Col. and Harding's King of Cols. and a number of his daughters were winners of the purple. Many other breeding sons were sent out by Morton into all section of the hog belt to improve Durocs with size, stretch and vigor. Among the most noted of these was King the Col., but recently dead, owned by Larson, of Iowa, for several years. At one time there were 48 sons of King the Col. at the head of Western herds, and his daughters have been good producers and are still much sought after to cross on top boars of the breed.

The get of old Crimson Wonder was in the meantime in the lead in Western herds and many prominent boars and producing sows became noted in this work. Among the best known progeny of old Crimson Wonder were Crimson Wonder I Am, mentioned in connection with the Iowa Championship, his son, Crimson Wonder Again (dam Orion bred). This latter boar crossed with H. A.'s Queen produced most of the noted Crimson Wonder·boars, such as Belle's Crimson Wonder and Crimson Wonder III., I Am Perfection Wonder, Nebraska Wonder, Uneeda Crimson Wonder, Long Wonder, I Am a Crimson Wonder, Crimson of Crimson Wonders, Red Chief Wonder, Allen's Wonder, B. & C.'s Crimson Wonder, Belle's Wonder, Champ Crow and others.

Crimson Wonder III, after winning champion at Nebraska Fair, 1908, and being used some time, was brought to Illinois by Browning & Comer, and when campaigned again won 1st in aged class in three of our largest fairs. He was a medium sized hog, and crossed only fairly well with the Colonels and Protections used in middle section and Eastern herds. His daughters were used in many herds east of the Mississippi, but only a very few of his sons proved valuable herd sires.

Another family which deserves notice here is the Critics. Glendale Critic, the head of this family, was bred by E. Z. Russell, and traces through Tolstoi, and Hugh Tucker to Protection, his dam being an Ohio Anna sow, one of the same family from which came old Crimson Wonder. Malcolm Model, by Old Top Notcher, was also used in building this family, as was Gold Dust Jim by Liberty Jim, the latter from the Roberts herd and tracing to Allison 5267. He had another boar also called Am Awake by Awake, a Roberts boar mentioned heretofore. Jumbo Critic, Glendale Critic, Jumbo Critic Jr., Critics Echo and other boars of this line were used to further intensify this blood. Later Mr. Russell used Proud Chief by Ohio Chief, injecting another line of Protection blood in the herd. The Critic boars were shown at Nebraska and Iowa State Fairs in their early years by Mr. Russell and won many high honors. Widle & Sons, of Nebraska, and other breeders of that State have carried the same blood lines to the big shows and won coveted prizes. It is worthy of note that the Grand Champion sows of both National Swine Shows held so far (1916 and 1917) were Critic bred sows.

One of the older herds of Durocs today and one of note for the line of high class sows produced therein, as well as a number of boars developed, and one that has been known in the leading shows of the East for a longer term of years than any other, is the Mahan herd, founded by Mr. Mahan, Sr., early in the history of the breed's improvement, and continued by his sons, C. E. and Pearl. Nellie Warren, little sister to Lucy Wonder, of Columbian Exposition purple honors, was one of the foundation sows. She was by King Keever by Ohio Bob, he by Rosemond's Royal from the Stoner herd. Three sows from the Abernathy herd tracing to the Vernal sows of the Morton herd were added early in the herd's history. King

Hustler, a boar from the Morton herd and Bon Ton, from the same source, were boars in early use. Duchess Czarina and Duchess Maud, sows by old Top Notcher out of Duchess Gem, she from the O. Walter herd, were foundation blood. One of the Protection boars they used was High Chief by Chief of Ohio and they also had Orion Boy by Orion II., also Top Chief by Winchester Chief. Later King to Be by Morton's King, out of a Protection dam was used and Good Choice by Choice Goods by Chief of Ohio by Protection. From King to Be they produced King Victor and Maud Irene, well known state fair winners. Mr. Mahan, Sr., took Champion boar prize in the first State Fair in Ohio with a boar of his growing. King's Lad, litter mate to King Victor was also used. Later they had another Ohio Chief bred boar, Grand Success, he by S. S. Success by S. S. Col. by Ohio Chief. Next came Kruger's Corrector by Kruger, of St. Louis Fair fame, and Chief to Be by Grand Chief by Ohio Chief. In later years they bought Fancy Col. in the Morton sale, he by King of Cols. II., by Old King of Cols., and then Pals Cols. by Premier Col., by King of Cols., and The Chief Col. by Cherry Chief,

Orion Col. M. by The Chief Col., out of a daughter of Orion Chief and Florazel, the noted show sow by Choice Goods.

The sows in this herd have always won high places in our best shows and from this herd many producing sows as well as show sows went into all parts of the hog belt. Mahan's have believed that no boar is good enough to produce the kind of hogs they wanted to grow by being crossed on ordinary sows, and while they have produced, grown and exhibited many famous boars of the breed, they made it possible by paying unusual attention to selecting only the very best sows to keep in their herd to produce this kind of get. The Duroc breed has suffered for the want of more breeders to carry out this same idea.

After the death of King of Cols., Morton placed Premier Col., a son, at the head of his herd, and he proved to be a good producer, producing such hogs as Volunter, Pal's Col. and others, and his daughters have produced many good Durocs. Later he bought from Johnson & Son, of Indiana, Morton's Top Col., a line-bred boar, being by Top Col., out of an Orion Chief dam, Top Col., his sire, being by

King of Cols., out of an Orion Chief dam. This boar died at the height of his breeding age, leaving several get that proved him to be a sire of rare merit, among these being American Top Col., still owned in the Barker herd of Indiana; Perfect Top Col., used as herd header by Morton, and later sold to Truax; Walt's Top Col., the latter used a good many years in the Jackson herd, and on the producing value of Walt's Top Col. daughters, much of the fame of old Orion Cherry King is founded. Walt's Top Col. was out of a dam by Royal Col., he by King of Cols., making him an intensely Colonel bred boar. He is still in use, being in the Wenger & Studebaker herds in Ohio; Taxpayer was another Morton's Top Col. boar and was used by Watt & Foust, sire of Taxpayer Thirteenth, Taxpayer's Model and others.

Orion Chief headed the Jackson herd for several years after the St. Louis Show. Kruger was also used in the herd. Orion Chief was sold to Thos. Johnson at $3,500, the record price at that time for a boar at private sale. He was later sold in Johnson's dispersion sale at 9 years of age, to Messrs. Matern & Mumford for $500, and died on the Matern farm. Jackson purchased Cherry King, a Protection bred boar, of Morton & Stewart Bros. in 1909, and founded the famous Orion Cherry Kings by crossing him on Orion Chief sows, and sows by sons and grandsons of Orion Chief. However, before this blood line became known by that name he sold his entire herd to Chas. Sprague, including Cherry King, Jack's Friend, by Joe Orion, by Orion Chief, and Joe Orion 2nd by Joe Orion, and retired from the public and private sale of Durocs by agreement for two years. Getting back into the active work after the elapse of the 24 months, he showed Orion Cherry King and won Grand Champion at Ohio in 1913, and from that date until the present the Orion Cherry King blood has been in the ascendency, taking more Grand Champion prizes than the Durocs of any other family of the breed ever did. Orion Cherry King's dam was Orion Lady A. by Orion Chief out of King Lady, she by Chief Surprise by Chief of Ohio a litter mate to Ohio Chief.

One of the most noted sons of Ohio Chief, produced after Mr. Harding took that noted hog West, was The Professor, a boar purchased as a pig by Henry Matern, of Illinois. He proved an unusually

great breeding hog, and sows bred to him
commanded the highest prices. He produced such boars as Instructor, Grand
Champion Illinois; Superba, The Principal 4th (at one time head of Mumford
herd), and a long line of good sows. When
the hog was five years old Mr. Matern refused $5,000 for him, the hog having made
his owner at least five times that amount
even at the low prices prevailing in those
days. He produced equally well on sows of
varied breeding, thus establishing the fact
that he was a real herd boar. Mr. Matern
won regularly with his get at Illinois and
the International for a number of years.

Another descendant of Ohio Chief that
came back from Mr. Harding to Illinois
was L. E.'s Valley Chief by Valley Chief
by Ohio Chief. He was owned in turn by
Sexsmith & Strong, Van Nice, L. E.
Thomas of Illinois. From him was produced L. E.'s Valley King, the sire of Illustrator, a boar that headed the Van Meter
herd and won grand champion Illinois and
sold to Dr. C. E. Still, of Missouri, for
$2,000, and in the latter's closing out sale
to Moats & Son, of Iowa. A litter brother,
Illustrator 2nd, went from Illinois and
headed the George Briggs herd. L. E.'s

Valley King won as a pig, a yearling and aged boar at Illinois, and was owned in turn by L. E. Thomas, J. Young Caldwell and Henry Matern.

For three or four years prior to the St. Louis World's Fair, the Browning herd, so prominent in early Duroc lore, was not on the show circuit, as increasing age forced the elder Browning to slack up his work with the breed and for a time Harry was connected with the Seckman herd. He helped develop and show Tip Top Notcher at the World's Fair, but soon after that time was again managing a herd of his own and the names Browning and Idlewild Farm became familiar to every Duroc man in America. Harry developed that great family of sows the Helen Blazes, and sold Helen Blazes III. for $1,000, a record price, and during the next ten years owned some of the most noted hogs the breed has ever known. He was a feeder, showman and builder of good Durocs and was a leader in shows and sales until 1915 when he closed out and has applied himself to newspaper work entirely. He was the first fieldman on the breed paper in the early days of that publication and did untold good in spreading

Duroc enthusiasm. He has always believed in Durocs, in gilt edged pedigree and in publicity. When King of Cols. came into the limelight, he and L. E. Thomas purchased Chief of Cols., one of the King's greatest sons, and developed him into a Grand Champion and he proved to be one of the best breeding boars of the Colonel family, producing sows with high backs and stretch with ability to produce good litters. Browning's herd was strong in Proud Advance blood lines. One of the most noted sows he ever owned was Lucy Wonder 21st, a daughter of Lucy Wonder of World's Fair fame (1893). Her litter, by Proud Advance, producing Proud Fancy, one of the most noted dams of the breed. Browning bought Defender as a yearling, fitted and showed him, won first at Iowa and Illinois and Grand Champion International, 1909, and Grand Champion International, 1910. In company with R. L. Comer he purchased Crimson Wonder III., a Nebraska show hog, and succeeded in intermingling this Western blood into his herd, which up to that time was of Colonel, Ohio Chief and Proud Advance breeding. The foundation of the Helen Blazes sows, Helen Blazes 64502n was shipped to Iowa and mated to Ohio Chief, giving him a Protection bred litter that added much to the popularity of the herd. Educator was one of this litter, and after being used by Browning and sold West, finally landed with the Fred Swan herd. Soon after selling Defender he purchased from S. E. Eakle the boar known as Disturber, he having sold the dam to Eakle & son, bred to Defender. This hog he fitted and campaigned and won many honors. Disturber was undoubtedly the smoothest and largest big type Duroc boar produced up to that time. He died soon after the close of the fair circuit. Disturber's dam was Lucy Wonder 112th, tracing to Chief's Col., Proud Advance and old Lucy Wonder. Browning developed the Tattletale family of sows and carried a number of them to leading shows. These came from the great line of brood sows Browning had produced in his herd and many of them were sired by Volunteer, a boar by Premier Col., by King of Cols., which Morton had sold to Dr. Stanberry of Tennessee as a pig. Stanberry developed and fitted Volunteer and won grand championship at Nashville, Tenn., 1911. Browning bought him for $1,000, won grand championship at the International, 1911, selling him to J. W. Storm of Ohio later for $1,500, who in turn sold the boar and his entire herd of sows to Thos. Johnson, March, 1913. Browning developed the public sale business in his breeding of Durocs to a science and held more noted auctions than any other Duroc breeder, the most noted being his sale of show and breeding boars in November, 1911, when he made an average of $303.65, something undreamed of in averages at that date. Eleven consecutive sales held by him in which he sold about 550 head made a grand average of $125.

One of the sons of Ohio Chief that went West before Harding & Osborn bought the old hog was Model Chief. This hog was bought from Morton by Watt & Foust and developed by them. As a yearling he sold to J. Coy Roach, the man who afterwards paid $1,000 for Helen Blazes III, the first sow of the breed to sell at that price. Through the efforts of the writer, Wm. Reed of Rose Hill, Iowa, purchased this boar of Roach and produced many hogs that have made history for the breed, one of which was Model Chief 2nd, Grand Champion of Iowa, 1908. Through the excellent line of sows produced by Model

Chief and his sons, much Protection blood was injected into the Crimson Wonders and Golden Models. Among the other producing boars used in the Wm. Reed herd was Reed's Banker, a half brother to Proud Advance, out of a dam by Malcolm Chief, by Jumbo Red by Protection. Following the work of Model Chief 2nd, Wm. Reed purchased Chief Select from the Mahan herd, he by Cherry Chief and later added Reed's Top Col., a son of Top Col. A son of Orion Cherry King was used by him the past year and a half, a boar now owned by Kerns of Nebraska named Great Orion. Mr. Reed's herd is not a large herd, but it is doubtless the most uniform type herd in the United States, and as a real builder of good hogs — a constructive breeder — he stands with few equals and no superior.

One of the herds which has come into the greatest prominence in the past five years is the herd of Prof. H. W. Mumford of Ann Arbor, Mich. He has been breeding Durocs almost as long as our oldest pioneer breeders, but was content to breed and build in his quiet unostentious way until he had the foundation deep enough and broad enough to come before the public with a herd worthy of his position and his ambitions. Owing to his position in the Illinois University keeping him at a distance from his herd, the building has necessarily been slower than if he had devoted his entire time to the work. Away back in the time of the early records he used a boar by Red Jacket from the Talmadge herd and the first recorded sow he owned was Plissy J., from the Stonebraker herd of Illinois. Ed's Eclipse, a boar from the Walter's herd in Ohio and a Col. M. boar and Liberty from the Morton herd were later used. Then two boars by Kant Be Beat from the Watt and Foust herd, one of them out of a Variety bred sow and one out of Rubertha, a winner in many fairs. This latter boar was called Fearnaught. Later a boar by Golden Rule was added, and Ohio Chief Again, the latter by Good Enuff by Golden Rule; King of Illinois from the McFarland herd was used a year or two. Then another Protection bred boar, Cherry King, Jr., by Cherry King, from the Jackson herd. This blood he intensified by later buying Brookwater Cherry King by the same sire. A few years ago he added more Protection blood from another source by purchasing The Principal 4th by The Professor by Ohio Chief, this boar's dam being Col. Stoner, he by King of Cols., and his second dam tracing to old Sensation 7393. Panama Special, an intensely bred Colonel boar, is now being used along with his Protection bred boars and sows. There is probably more Protection blood in Mr. Mumford's herd than can be found in any herd in the land. Prof. Mumford is one of our greatest Duroc judges and his work in the 1916 National Swine Show did more to unify Duroc breeders on proper type than any event of the recent past.

Duroc history in the East would be incomplete without a notice of the Johnson & Son herd of Indiana. Among their first noted boars was John's Ohio Chief. This boar was out of Nellie Morton, she sired by Morton's King. John's Ohio Chief's daughters were probably the best producing sows the East has ever known. They were not a large type, but were smooth, deep bodied, mellow and easy feeding. Mated with Colonel boars, they have made much history in shows and sales. Later Johnson & Son purchased Chief's Top Lady from Ira Jackson. This sow had been shipped to Mortons and mated to King of Cols., at the same time Lena J., dam of

Defender, was mated to this noted boar. From this litter came Top Col. He was used until the herd was dispersed, when he sold to Truax & Son of Ohio at $1,000. The Messrs. Johnson also produced Cherry Chief and sold him as an under year boar to Morton & Co., and the old hog later was purchased by J. C. Droz, Hanks & Bishop.

Ernest Pancake of Illinois assembled a herd soon after the World's Fair in St. Louis and had at the head Ransom Chief, by Ohio Chief, one of the litter mentioned as produced by Browning, from old Helen Blazes and Ohio Chief. He also bought Prince Wonder, a boar by Decatur Boy of Jackson. This was one of the most perfect individual boars the breed has ever produced. He died while being fitted for Illinois. Prince Wonder sows were all good producers. One of his sons, Prince Wonder Again, was grand champion at Minnesota, 1911. Pancake later purchased Defender and in 1911 sold him and his entire herd to East Bros. of Ohio.

One of the families that has come into prominence in the last three years, and which seems destined to be in the lead for some time, is the Sensation family. At the present writing more boars of this family are before the public in the West than the boars of all other lines combined. A large majority of these boars have been bred and developed by Wm. Moderow of Nebraska, a hog man of the first rank. The name Sensation is not new for we find back in the early years of the breed the old boar Sensation 7393, sired by Orion R. by old Orion, Orion R. being farrowed in the Robert's herd and sold to Moorman of Indiana, one of Indianas pioneer breeders. Sensation was out of Hoosier Girl, a daughter of Winchester Chief, and the latter is out of old Duchess 40th, the dam of Ohio Chief and Top Notcher. Sam Stewart of Nebraska secured a son of Sensation from D. W. Brown of Indiana who had purchased old Sensation from Moorman and used him in his herd a number of years. This young boar, called Chief Sensation, was out of a sow that came from the Trone herd in Illinois and whose lines trace to Trones Hero, a winner at Chicago, 1893. One of the noted sows produced in the Stewart herd from this boar was Sensation Girl, her dam a great granddaughter of old Top Notcher and Gold Dust Jim, the latter a Roberts and Russell bred boar that produced much improvement in Nebraska and Iowa Durocs in his time. Sensation Wonder and Sensation Wonder II came from a mating of Sensation Girl, with Wallace's Wonder, one of the many noted Crimson Wonder boars that came out of the cross of H. A.'s Queen with Crimson Wonder Again. Sensation Rose, the dam of Great Sensation (Kern), King Sensation (Labert), and Top Sensation (Moderow & Toelle) is by Sensation Wonder, out of Red Rose, carrying a combination of Protection, Orion and Kruger breeding. The name Sensation has been used with all the prominent boars of that line, although they are as much Crimson Wonder as they are Sensation bred. Wallace's Wonder was produced by Wm. Sells from the Henry Allen boar, Crimson Wonder Again. Clarence Wallace produced Sensation Wonder from Wallace's Wonder and Sensation Girl. Sensation Rose was mated by Moderow to Great Wonder I Am and produced the three Sensation boars mentioned in the first of this paragraph as well as many others of less note and some excellent sow stuff. Great Wonder I Am carries eight crosses of Protection breeding, seven of Crimson Wonder, two of Golden Model blood as well as a touch of

Orion and Critic lines. Great Wonder I Am's top lines trace back as follows: Great Wonder I Am, Great Wonder, B. & G.'s Wonder, P. V.'s Wonder, Lincoln Wonder, Ohio Chief and Protection, the name "Wonder" in this line coming from the dam of Lincoln Wonder, Morrison's Peach, a daughter of old Crimson Wonder, a noted sow of her day. The family of Sensations came prominently before the general public at the 1915 Nebraska Fair, when under year boars by Sensation Wonder 2nd won 1st, 2nd, 4th and Junior Championship at the show. Since that time the ascendancy of this line has been rapid and some of the largest boars and some of the best sows of the breed carry this line.

The line of breeding known since 1912 as the Defenders is really an intensifying of the Orion and Colonel blood. This Defender blood has been linebred and inbred by the McKee Bros., who purchased old Defender in 1913 and have continued to improve and perpetuate this line. They have proven themselves real swine builders, and have produced a type peculiar to that blood line and a type that conforms in every particular to the large, stretchy, high-backed hog which all breeders are striving to produce. They had in their herd as foundation material, sows of Ohio Chief, Golden Rule, Proud Advance, Orion Chief and Colonel lines, and produced many noted boars by these matings, but are now mating Defender boars to their Defender bred sows and have continually increased the size of their breeding animals, retained their prolificacy and produced one of the most prepotent lines of breeding which the breed has ever known. The Defenders are more intensely bred than any other family of the breed. Possibly the best boars and sows produced in this herd have been from the cross of Defender blood, with granddaughters of Cherry King, although the writer has contended that one of the best breeding sons of old Defender was Pilot Defender, his dam by Kelly's Pilot Wonder, a grand champion of Ohio several years ago.

Another of the boar progeny of the Crimson Wonders that came into prominence is old Model Wonder 53981 by Crimson Wodner I Am, out of a dam tracing to Pericles of the Roberts' herd. Pericles by old Orion. Hanks & Bishop bought this Wonder boar from U. G. Davidson and used

him in their herd following Top Notcher
Again by old Top Notcher and other boars.
From Browning they purchased Proud
Col., a boar sired by Chief's Col., and out
of old Proud Fancy by Proud Advance,
and produced a wonderful lot of big,
smooth brood sows. Later they purchased
the old sire, Cherry Chief by High Chief, by
Ohio Chief, and a few years ago followed
with Pathfinder, one of the largest boars of
the breed, a boar carrying the blood of Protection and the Colonels, his sire being Pro-
phetstown Chief by Cherry Chief by High
Chief by Ohio Chief by Protection. His
dam is by S. E.'s Premier Col., by Premier
Col. by King of Cols., his second dam by
Orion Chief.

We have, several times in the history,
referred to Cherry Chief 21335 and his
progeny play an important figure in modern Duroc history. He was sired by High
Chief, he by Ohio Chief and was out of
Cherry Queen, she by Baker's Bred Right,
the latter boar carrying the same blood
lines as old Orion. Johnson & Son of Indiana bought Cherry Queen and afterwards had her bred to High Chief. In No-
ember, 1907, Morton & Co., a firm composed of S. E. Morton & Stewart Bros.,

bought him and had him on their farm until he was six or seven years of age, when they sold him to J. C. Droz of Iowa, who in turn sold him to Hanks & Bishop. Among noted hogs sired by him were Chery King, one of the foundations of the Orion ·Cherry Kings; Cherry Chief II, owned a number of years by Thos. Johnson of Ohio; The Chief Col., at head of Mahan herd and later sold to Bartley and to Gerlaugh; Chief Select, a boar sold by Mahan into Iowa, and used in the Reed herd and other prominent herds in that state.

One of the Duroc blood lines that was on the top round of popularity for a time, and a strain that showed as much easy feeding qualities as any the breed ever developed, was the Golden Rules, perfected and produced in the Watt & Foust herd and carried by them to many high honors in the Ohio State and other shows. Golden Rule was sired by Choice Goods, a Protection bred boar in the Jackson herd, the dam being Mayflower, a sow tracing to Col. M., the foundation of the Colonel family. Golden Rule produced Choice Rule and Good Enuff, both Grand Champions at Ohio, and Good Enuff produced Good Enuff Again, also Champion at the same state fair in 1910. The latter was sold to W. H. Robbins of Ohio for $1,000 and this boar weighed 925 pounds at 17 months of age. Good Enuff Again produced Burke's Good Enuff, a winner as a pig at International and was sold to C. F. Burke of Colorado. He won grand championship at Colorado and afterwards was sold to Economy Stock Farm, Iowa, and won for them grand champion at Illinois, 1915, and was used extensively in that herd. Good Enuff was also sire of Pride Enuff, used in the McKee herd a short time. The blood line of these boars, produced by Golden Rule, was founded on the noted Cedarvale sows in the Watt & Foust herd.

In the McFarland herd of Missouri the name Wonder has been extensively used. This does not come from the Crimson Wonder name, but from the fact that they purchased some years ago descendants of old Lucy Wonder, Grand Champion at Chicago, 1893. They purchased at one time six gilts of this breeding from O. Walter & Bro., that were sired by Longfellow, Jr., by Old Longfellow 6815, and were out of a grand daughter of Lucy Wonder. They also used Oom Paul from the Trone herd, a grandson of Jumbo Red.

Also Gold Finch, a boar with same blood lines as old Orion. They won grand championship at St. Louis with Dotie, a sow they had purchased of C. R. Doty and the following year had her mated to Tip Top Notcher and since that time boars of this latter breeding have been used in their herd.

SOME CHAMPIONS OF THE PAST

B. & C.'s Col. 80587n (Col. Carl 19265a) was full brother to King of Cols. He was grand champion of Iowa and Illinois, 1909, shown by Baxter & Comer.

Rosemary Duchess, by King of Cols., champion sow of Ohio, 1908. Sired by King of Cols., and was owned and shown by Whitehall Farm, Ohio.

Col. S., by King of Cols., was grand champion, Kentucky, 1908, shown by Whitehall Farm.

Much Col., by Chief of Cols., he by King of Cols., champion, Indiana, 1909, shown by D. W. Brown.

Miss Orion, by Orion Chief, champion, Indiana, 1909, shown by Mahan.

Medoc, Jr., grand champion, Nebraska, 1909, and grandson of Kantbebeat, out of a Reed's dam, shown by Van Patten.

Golden Queen, grand champion sow, Nebraska and Sioux City, 1909; shown by Waltemeyer.

King of Col's. Ideal, by King of Cols., grand champion, Ohio, 1909; shown by Cline.

Duchess Czarina 4th, grand champion, Ohio, 1909, sired by Top Chief, by Winchester Chief and out of a dam by Choice Goods, by Chief of Ohio, litter brother to Ohio Chief, second dam by old Top Notcher, shown by Mahan.

Crimson Jewell, grand champion sow, Iowa, 1909, sired by Ohio Chief, out of a Proud Advance dam; shown by Hanks & Bishop.

Freed's Col., grand champion, Iowa, 1910, sired by Prince of Cols., shown by Freed and afterwards owned by Freed & Harding.

Golden Queen 3d, grand champion sow, Iowa, 1910, sired by Golden Model 2nd, out of a dam by Model Chief, by Ohio Chief; shown by Waltemeyer.

Belle's Crimson Wonder, grand champion boar, Nebraska, 1910, sired by Crimson Wonder Again, out of a dam by Savannah Belle's Chief, by Ohio Chief; shown by Barnes.

Jack's Friend, by Joe Orion, by Orion Chief, grand champion, Ohio, 1911. Dam by King's Pal, by King to Be; shown by Jackson.

Pal's Prince, grand champion, Indiana, 1911, sired by King Pal's Last, by King's Pal, dam by Prince of Cols.; shown by

Harvey and sold to Sprague of Ohio for $1,010.

Colonade, full brother to Pal's Prince, grand champion, Indiana and International, 1912. Shown at Indiana by Harvey and sold to Goodwin, Illinois, for $900 and shown by Goodwin at International.

Mc's Dream, grand champion sow, International, 1911, and Wisconsin, 1912, Illinois, 1913, sired by Fancy Orion Chief by Orion Chief; shown by Browning.

Mo. Model Top, a grand champion Missouri, 1911, sired by Model Top, by Golden Model; shown by Sheley & Clatterbuck.

High Model by Golden Model 2nd, out of an Ohio Chief Dam, grand champion, Iowa and South Dakota, 1912; shown by Waltemeyer and afterwards sold to Shanks, Minnesota.

Chief's Maid by Valley Chief, by Ohio Chief, out of a Top Notcher Again dam, grand champion sow at Iowa, 1912; shown by Davis.

Col's Pilot Wonder, by Kelly Pilot Wonder, by Brock's Wonder, by Pilot Wonder, grand champion, Ohio, 1912; shown by McLaughlin and sold to Smith & Rogers for $1,250.

Critic B., by Dusty Critic, by Glendale

Critic, grand champion, Nebraska and Kansas, 1912; shown by Danford and later owned by Widle.

Sunbeam Girl, by B. & N. Chief, dam by Neb. Critic, grand champion sow, Nebraska, 1912; shown by Hansen.

Valley King, by Valley Chief, by Ohio Chief dam, by Top Notcher Again, grand champion, Iowa and Nebraska, 1911; shown by Harding.

Orion Cherry King, grand champion, 1913, shown by Jackson.

Fancy Flo, by Fancy Col., grand champion, Ohio and Kentucky, 1913; shown by Mahan.

The Chief's Model, by Cherry Chief, grand champion, Indiana and Kentucky, 1913; shown by Mahan.

Big Wonder, by I Am Crimson Wonder, out of a Kruger bred dam, grand champion, Iowa, 1913; shown by Stevens and sold to Economy Farm.

Golden Model 34th, by Golden Model 2nd, grand champion, Nebraska, 1913.

Fancy Advance 2nd, by Wallace's Wonder, by Crimson Wonder Again, grand champion sow, Nebraska, 1913, shown by Wallace.

Illustrator, by L. E.'s Valley King,

grand champion, Illinois, 1913; shown by VanMeter.

Lady Climax by II Climax out of Dotie, the St. Louis grand champion sow, grand champion, Missouri, 1913; shown by Mc-Farlands.

Nebraska Belle, grand champion, Nebraska, 1904, sold to Harding for $600, a record sow price at that time.

Joe Orion 2nd, grand champion, International, 1913, by Joe Orion, by Orion Chief; sold by Sprague to Enoch's Farm at eight years of age at $5,000.

Grand Model, grand champion, Iowa, 1914; shown by Waltemeyer and retained in their herd as herd header until his death.

Golden Queen 35th, grand champion sow, Iowa, 1914; shown by Waltemeyer.

Echo's Crimson Wonder, grand champion, Nebraska, 1914; sired by Van's Crimson Wonder, by Crimson Again dam, by Echo's Top King.

Imperator, grand champion, Kentucky, 1914, sired by Fancy Col., by King of Cols. II, by King of Cols; dam by Col. J., by Tippy Col., by Prince of Cols.; shown by Williams, sold to Mayfield Farms.

Royal King by Orion Cherry King, grand champion, Ohio, 1914; shown by

Jackson and sold to Johnson Bros., Minnesota, for $2,650.00.

Critic D., by Critic B., grand champion, Nebraska, 1915; shown by Widle.

Grand Lady, by Grand Model, grand champion, Nebraska, 1915; shown by Waltemeyer.

Golden Queen 16th, by I Am Golden Model 2nd, grand champion, Iowa, 1915; shown by Waltemeyer.

Model Select 2nd, by Model Select, by Chief Select, by Cherry Chief, grand champion, Iowa, 1915; shown by Spencer.

Gold Certificate, by Gold Bond, by Good Enuff Again, grand champion boar, Ohio, 1915; shown by Robbins.

Pal's Lady, by Pal's Col., grand champion, Ohio, 1915; shown by Gerlaugh.

Taxpayer Thirteen, by Taxpayer, by Morton's Top Col., grand champion at San Francisco, 1915.

Crimson Elizabeth, grand champion sow, San Francisco Exposition, sired by I Am Perfection Wonder, by Crimson Wonder Again; shown by Hoover.

Taxpayer Model, grand champion, Atlanta and other southern shows, 1917, sired by Taxpayer by Morton's Top Col., bred by Watt & Foust, winner as pig at Ohio;

sold to Mahan Bros. at public auction at $550, record price for a six months' pig; sold in Mahan closing sale to Coldstream Farms.

Orion Cherry King, Jr., grand champion, Ohio and National Swine Show, 1916, sired by Orion Cherry King, dam by Jack's Friend; shown by Jackson & Foust, sold to Peacock & Hodge, Georgia.

Joe Orion King ("Scissors"), grand champion, National Swine Show, 1917, sired by Orion Cherry King, out of dam by Joe Orion II; shown by Jackson, sold to Pine Crest Farms.

Fancy Orion M. 2nd 277198

Grand Champion Duroc Jersey Sow, Ohio State Fair, 1918

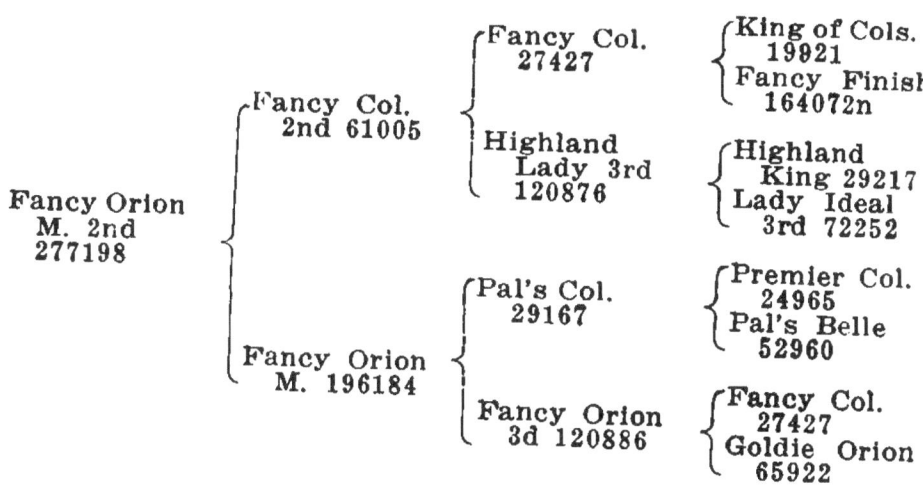

Perfect Defender

Grand Champion Duroc Jersey Boar, Indiana State Fair, 1918

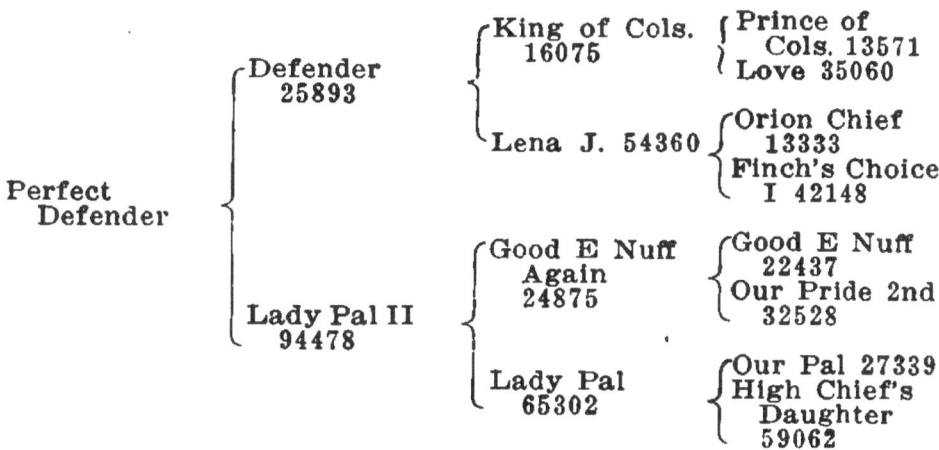

Grand Lady 66th 567778n

Grand Champion Duroc Jersey Sow, Iowa State Fair, 1918

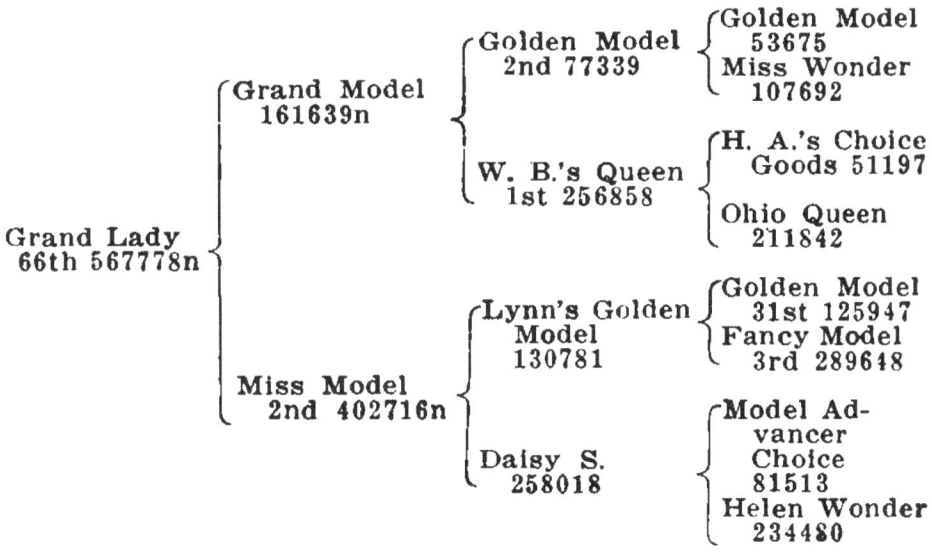

Pathfinder's Likeness 220343n

Grand Champion Duroc Jersey Boar, Iowa State Fair, 1918

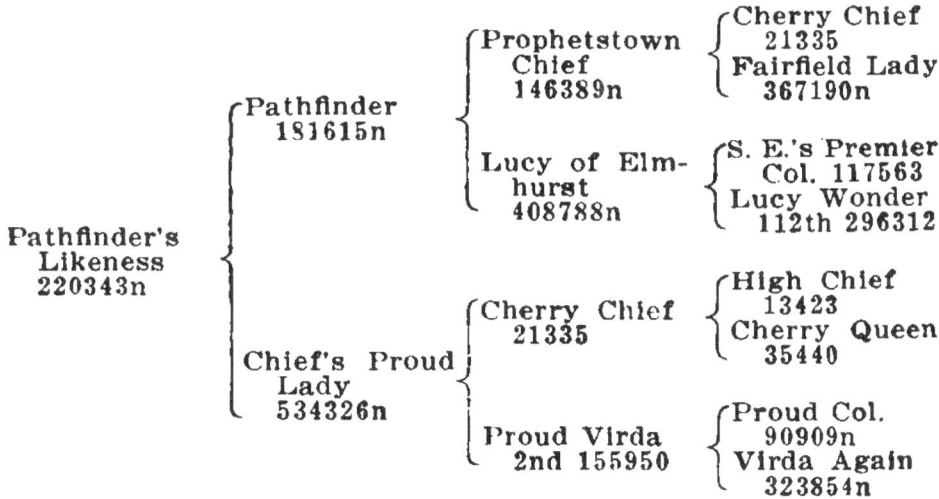

Colletta 3rd 179582

Grand Champion Duroc Jersey Sow at Southeastern Fair,
Atlanta, 1918

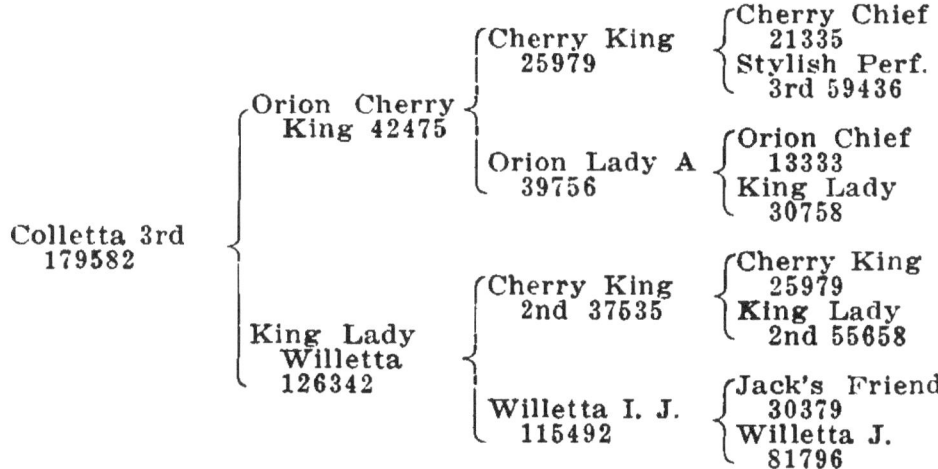

Colletta 3rd
179582
  — Orion Cherry King 42475
    — Cherry King 25979
      — Cherry Chief 21335
      — Stylish Perf. 3rd 59436
    — Orion Lady A 39756
      — Orion Chief 13333
      — King Lady 30758
  — King Lady Willetta 126342
    — Cherry King 2nd 37535
      — Cherry King 25979
      — King Lady 2nd 55658
    — Willetta I. J. 115492
      — Jack's Friend 30379
      — Willetta J. 81796

Taxpayer's Model 56529

Grand Champion Duroc Jersey Boar at the Southeastern Fair. Atlanta, 1918

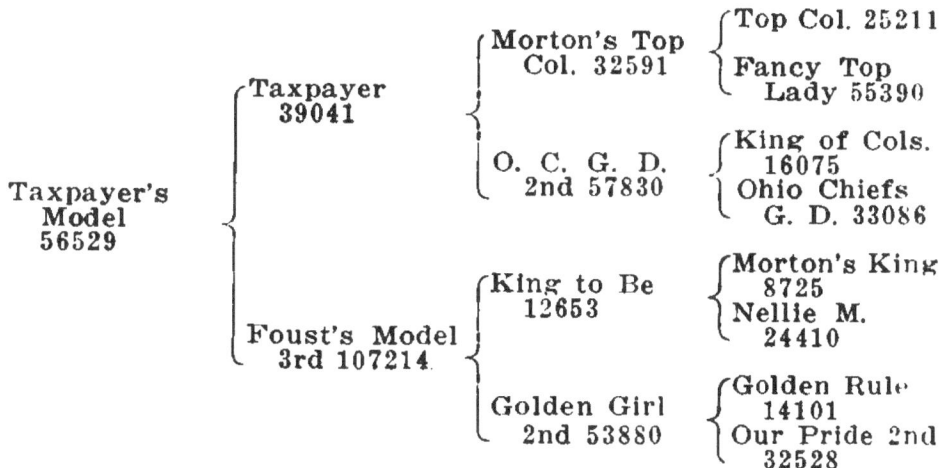

Taxpayer's Model 56529
- Taxpayer 39041
  - Morton's Top Col. 32591
    - Top Col. 25211
    - Fancy Top Lady 55390
  - O. C. G. D. 2nd 57830
    - King of Cols. 16075
    - Ohio Chiefs G. D. 33086
- Foust's Model 3rd 107214
  - King to Be 12653
    - Morton's King 8725
    - Nellie M. 24410
  - Golden Girl 2nd 53880
    - Golden Rule 14101
    - Our Pride 2nd 32528

Long Gano 536804n

Grand Champion Duroc Jersey Sow at the Nebraska State Fair and the National Swine Show, 1918

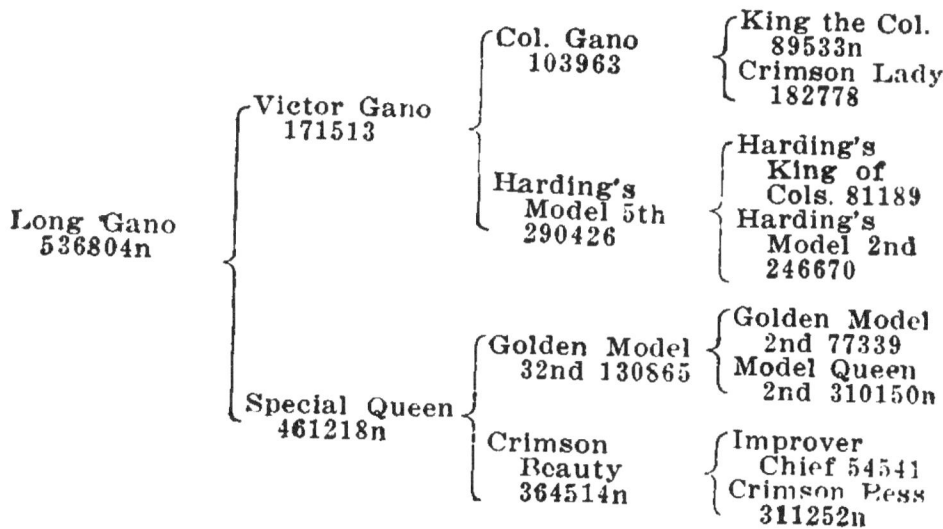

Great Orion 128377

**Grand Champion Duroc Jersey Boar at the Nebraska State Fair
and the National Swine Show, 1918**

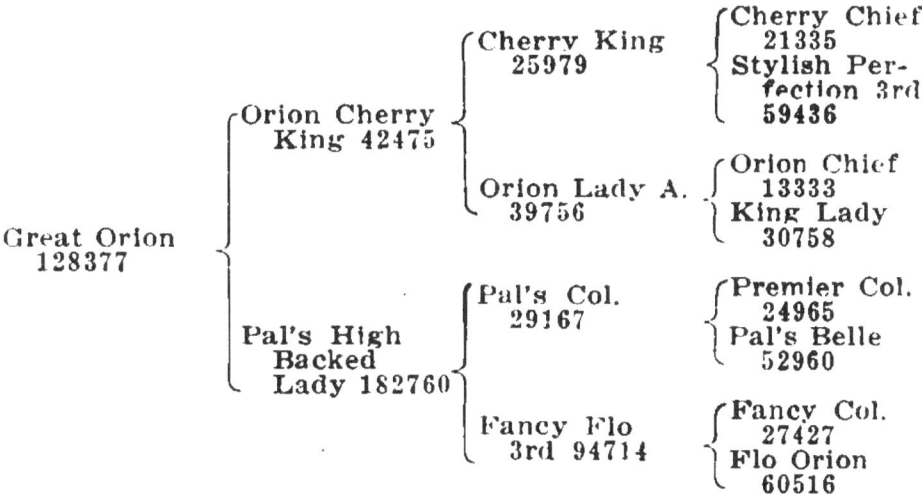

Throughout this book you will find a few photos of Grand
Champions and their extended pedigrees.